ENGLISH SUBT

ENGLISH SUBTITLES

PETER PORTER

Oxford New York Toronto Melbourne
OXFORD UNIVERSITY PRESS
1981

Oxford University Press, Walton Street, Oxford OX2 6DP
London Glasgow New York Toronto
Delhi Bombay Calcutta Madras Karachi
Kuala Lumpur Singapore Hong Kong Tokyo
Nairobi Dar es Salaam Cape Town
Melbourne Wellington
and associate companies in
Beirut Berlin Ibadan Mexico City

British Library Cataloguing in Publication Data
Porter, Peter
English subtitles.
I. Title
821 PR9619.3.P57 80-49705
ISBN 0-19-211942-7

Typeset in Great Britain by
King's English Typesetters Ltd., Cambridge
and printed by
Lowe & Brydone Printers Ltd., Thetford, Norfolk

For Anthony Thwaite

Acknowledgements

Acknowledgements are due to the editors of the following periodicals in which some of these poems first appeared: *The Age*, *Ambit*, *The Bulletin*, *Encounter*, *Honest Ulsterman*, *Nation Review*, *New Review*, *New Statesman*, *Poetry Australia*, *Poetry Book Society Christmas Supplement*, *Poetry Review*, *Times Literary Supplement*, *Thames Poetry*, *Quarto*, and *Vole*. Some others were first read in radio programmes broadcast by the BBC.

Once more I should like to acknowledge the help given me by the Literature Board of the Australia Council. The year's Fellowship which they granted me made the completion of this book possible.

Contents

English Subtitles

How much we need these annotations
When every frame prepared by the exacting eyes
Is a tableau from a famous series!
I can gaze from the eighth floor of the tower
Quite unconcernedly, but cannot avoid
Registering a fresco; then think of the intensity
When our looking is done within, the promontory
Of dreaming. It's all foreign languages,
Unfamiliar feelings, yet absolutely known
So that the craziest, most autonomous
Bulb-grower or Christmas fantasist
Is an unattainable part of the plot
Helping things make sense; or if rats
Coming ashore bring history with them
It will be hung round the necks of plants
In our well-cued crystal palaces.
And all the magazines I read,
Those articles which make me reach for
My spectacles spotting the word pluralism,
Their substance is our beautiful Northern tongue,
So useful for asking for fasteners in
As well as for caning professors:
I suppose my feeling foreign is just because
I have never got used to ambition
In my friends – how can they think it worthwhile
Dazzling and adoring so persistently,
Waking up to find themselves half-famous?
Although I feel I am becoming extinct,
A sort of adaptation to unloveableness,
I also have my calling, the quest for
Freedom wide enough to die in. Thus spake the words,
Just as I was swearing to abjure them all:
'Perhaps you should say something
A bit more interesting than what you mean.'
O bold deliverance! I can face my books

And the well-lit feast of journals,
I am anthropology itself. The used-up world
Becomes my kindly uncle: we are going
To that fascinating film which is always showing
To the cognoscenti. As we sit in the dark,
I turn the more uncomfortable passions
Into slim sentences, such old words
As I know you know. This is language
I would go into the jungle with,
English of the Rider Haggard sort
Reversed-out on the glare of consequence.

The Winter Capital

Sinking with distracted plumage now
Venus' doves decline towards the hospice
Of the West. The corrupted sea looks up
At keels furnished for eager roads
And tall economies. Blazon us
Upon the air, sigh defeated magistrates,
We should have stayed with princes
Among flat keys, a lengthy footage
Of worn wars and unimpressive foreheads.

Tread on the earth with the sweet
Misura of a perfect instep – it is not
Too late to be free of history:
Sisters of the quail in broad sun hats,
You have the Cart of Venus under you
And plural cities waiting. No one
May break the dance and changing partners
Is called love. Not just Winter
But the genius of the world has turned.

The Need for Foreplay

Is felt by these trees
accepting the wind along
their winter arms.

By dogs sniffing species
upon the thawing path
or stalled at fashionable ankles.

Even by the sun itself
greeting and then retreating
behind cloud for further teasing.

Earth with its motor defect
lurches suddenly to Spring,
over-explaining, under-achieving.

We are out early too,
unrepentently interpreting
natural by learned responses.

Amazed sophistication
to love the mechanics of love,
go backwards into mystery.

My Old Cat Dances

He has conceived of a Republic of Mice
and a door through the fire,
parables of the reinstatement
of his balls. But not this night.
Isn't there a storm in the light bulb,
condors circling the kittens' meals
on the television screen?
He heard once that people wearied of
each other to escape unhappiness.
In his lovely sufficiency
he will string up endless garlands
for the moon's deaf guardians.
Moving one paw out and yawning,
he closes his eyes. Everywhere
people are in despair. And he is dancing.

Visiting Cornish Churches

Folded in lush combe
Or sentinelling the land,
The wide-naved churches stand,
Like platitudes of doom.

Lanteglos with Polruan,
Lansallos by Lantivet –
Paths of the stormy privet
Where leaves soundlessly are strewn –

Hoarders of forgotten saints,
Sites for moral doggerel,
Lit by the improbable
Gold of restorers' paints,

Sanctuaries of afternoon
When the sun lies in the wheat
And flower-arrangers meet
To make comfortable God's room.

Past Celtic Cross and over
Graves at a hundred angles,
Through grass and nettle tangles,
The tourist breaks from cover.

The air he breathes is clean
And roseate with death,
Pevsner-listed souls beneath
Share with pew and screen

Small absolutes of fame:
Nothing remarkable here
But men's and women's fear
Of losing even a name,

And when he comes to quiz it
No monument will keep
Him long. He hopes they sleep
The better for his visit.

How Important is Sex?

Not very. Even if it plays a not
Inconsiderable part in misery,
You can be unhappy without reference
To its intervention or its absence.

Our researchers have discovered even
Species whose reproductive processes
Are quite unsexual – and usually these
Are the more efficient and uncomplicated.

But, says the man waiting for a letter
And trying to read an article in a liberated
Magazine, I haven't been able to keep
My mind off sex since I was seven.

Others' minds go further back. Perhaps
Our evolution took the one track
(As the mind has it) into love and found
That those innovatory machines

The genitals, once in place, wouldn't
Be denied their significance. The sight
Of mummy's hair puts us on the spot,
A cave more mysterious than the mouth.

Now flow from it plays and operas
And the horrible spoutings of rancid
Kitchens: a world of novels awaits
The boy taught things by his jokey schoolmates.

But you are talking about love, you'll say.
Yes, and I know the difference,
Taking down a wank magazine,
Then a note more fingered than any photo.

Nevertheless, I am a respecter
Of power, having seen a skinny girl
Screaming in the playground, oblivious
Of boys, wake to her hormonal clock

As Juliana or as Mélisande –
Even the great gods and captains
Might relax with a plaything
As bold and changeable as this.

The American Articulate

The opening-up,
audacity of
thinking what terrifies
with its demands
on heroics,
on the here and now –

Could it be voices
filling a new demesne,
gossip in the stockade
till the spirits stop ringing,
chirp, chirp of species
soon to be extinct?

Novelist by day,
chucker-out in a brothel by night,
the stoicism of prose
and the garrulity,
paradigms
of the making of money.

Knowing it will spread,
that unconscious fondness
for being first,
until the grieved globe
shrinks to Tennessee –

Far from aristocrats
keeping ahead of taste,
their meringue of gods
and cavalry lieutenants,
ready to follow
a trail to
the other ocean,
the big one of loneliness.

She wouldn't come downstairs,
he walked to the office
as if it were Königsberg,
their talking done on paper.

And now the flowers appear
on the earth. Nobody can be heard
above the shouting land.
What would they say?
What could they leave unsaid?

A Philosopher of Captions

The knowledge anyway is worth something,
That no person from this liner-browed brain
Will reach the height of those grave captains
Whose Dantesque walk and Homeric facing
Still flare on our desolate concrete plain
So late; that I am a philosopher of captions.

This special authenticity must grow on one
After baffled if dutiful years putting down
Some orders of words towards definition –
Here space a fear and there placate a pun,
Or adjudicate through childhood, one noun
Up and another down, with everything a fiction.

And the shouters, the ones met at stations by crowds,
One can only admire them, join the acclamation
And worry at their simplifying stance. The text,
After all, belongs to its explainers; those clouds
Are felt only as rain; an acceleration
In the speed of madness, harder saving from the wreck.

But the power is still somewhere in us, hovering
In the forehead auditorium of sounds:
Those who were with us and have changed their shape
Come back, like old ladies with parcels moving
To the chair beside us; embarrassment abounds
That pain is the one immortal gift of our stewardship.

Occam's Razor

Never take the more unlikely explanation
of any event in preference to the likely –
just so, I say to myself, as I consider
my life and conclude that this long haul
to a sort of maturity is nothing more
than a persistence of arriving in someone
with nowhere to go, rather than, as I once thought,
an inveterate but soon to be vindicated
earnest restlessness at the port of life,
a magic resistance to time, enough to
keep me the youngest person in the room
as well as the most ironically serious.

But then reflect that if one is to tease
out feelings from razors, one might prefer
to think of Ibsen's madman sharpening himself
or, in grateful suddenness, Haydn offering
his English visitor that marvellous quartet
in F Minor – what the gods mean by words
goes the long way round or takes on flesh
in dreams for the far vistas of nicknames.

Returning

Nobody feels well after his fortieth birthday
But the convalescence is touched by glory
So that history's truculent deeds of hate
Are lived through in dreams, the story
Followed to the investigator's hut, pain seen
Through a window on its knees, late
Help lost over marram dunes or never
Felt at the deliverance on a screen.

Marvellous means of escaping time and time's
Chosen people: sleep on the knowledge of
God's monsters! The school of love and crimes
Is open every night and the sedentary
Heir of men of action dashes off
A sonnet before execution. What you see
Is coiled in an uneventful past, rough
Justice of the body's failures, a commentary.

Yet never daring enough, even those hours
When the timid rule of truth relents
And every written word is without sense
As in some ultimate avant-gardish shape –
The apostle of plain dullness has powers
Of arrest and will use them; nightmares
Are prized categories too, a southern rape
Modelled in blood but with a classic tense.

It is time to recompose the face
Into a serious map, the children now
Envied creatures across a room, the case
Being settled for the present. Home is
The veteran of the adjectival run,
His images intact. He has learned how
To live another day and wakes, ringed
By the golden wallpaper of the sun.

Good Ghost, Gaunt Ghost

She is coming towards me,
looking at me to turn me to stone,
saying my name and turning herself
into territories I know from books,
into the damned who are behind blinds,
the peaceful madmen of the parish.

She has walked through an invisible screen
into the fire of every change,
a certificate of final adaptability –
she will dress in a novel
and loiter, as is usual, in a dream,
but that is accountability.

Her clothes are syntax, so that I read
someone else's poem and I am there
on the banks of salvation
or crying in a furnace. Why hast thou
held talent above my head
and let me see it, O my God?

Her shadow is rational, rationed of
tears and nocturnal commissions
saying the ego is always sublime,
the sublime always anticipatory,
and shadows our sisters under the skin:
each time we return to earth we die.

Words importing the masculine gender
include the feminine gender. Exactly,
and I see her as my hero-coward
who has dared to be myself, erasing
caution and suspicion. Soon I will be her
and we shall keep creation to ourselves.

Bei Einer Trauung

Off to the slaughterhouse
Of his or someone's imagination,
She has dipped in white
And orange blossom the tongue
Of her trust; a gambler
Raising the stakes, a cause
Of tension in the falling stockings
Of all old bags, she spools
Ordinary days from tangles
Of hope: whoever heard of
Liking at first sight?

O love, it must be love!
The face, careless of mirrors,
Will buzz about the house
Outraging cynic dust.
The organ hangs with glances
Which often raised her legs
In occasional beds,
But now she's comet-struck
By under-gods. Bury the father,
Snub the mother, club the lover.
Afternoon enters in its morning dress.

Sonata Form: The Australian Magpie

It makes a preliminary statement
with its head to one side and an eye
far too large to be seemly.

It is no relation to the English magpie
yet is decently black and white,
upstaging its cousin the kurrawong.

Its opening theme is predation.
What it scavenges is old cake
soaked in dew, but might be eyes.

Such alighting and strutting
across the mown grass of the Ladies' College!
Siege machines are rolling near.

Bustle in a baking tin,
a feast of burnt porridge –
the children are growing on their way to school.

You can upbraid the magpie,
saying, 'What do you know of Kant?'
It might shift a claw an inch or two.

It can tell when an overlord is unhappy.
When one sweeps out in tears to clatter
the petrol mower, magpie flies off.

But never flies far. Big feet
are moving to their place in dreams –
a little delay in the sun won't count.

We have certainly heard this theme before,
the sound of homecoming. Anticipation
needs a roof, plus a verandah for magpies.

Are these the cries of love or of magpies
sighting food? Some things about desire
call for explicit modern novels.

Magpie talk: Nation, National, Nationalist!
In this tongue its name is legion.
We speak English ourselves, with a glossary.

The coda, alas. It can be Brucknerian.
We say the end is coming. The magpie
has found its picture in an encyclopaedia.

Where can there be nature enough
to do without art? In despair, the poet
flies to the top of a camphor laurel.

Girl and magpie leave him in the tree.
Tomorrow a trip down the coast for her
and spaghetti rings left out for the bird.

All the Difference in the World

Between the sun beyond the window
and the black Latomia of despair
where her picture lives and where
sexual spasm jets into the ear of Dionysus.

Between those friends on the stairs
bringing wine and praises and the one voice
at midnight reminding me that reparations
are exacted of the talentless.

Between the memories of floral sheets
exuding the smell of us in bed
and the hospital ending in
a vapour of morphia and cocoa.

Between the Heavenly Philharmonic
and a sea raging at an inner port
whose veins spell death by pressure,
a stroke on the tympanum.

Between the arrival of letters,
neither long enough nor sufficiently tender,
and the platitudes of dreams
stalled on their evolutionary ladder.

Between wounds made by words
and the enduring silence of those
who can talk of love
only in the cadences of memory.

Between the soul as silkworm,
spinning out its time for a new house,
and the soul as a blackened Strozzi
churched in death and unawakenable.

Between poems which make litanies
of our being born and dying
and oracles which mislead us
about the differences between.

Myopia

It is to see things as God sees them,
up close, thread on thread,
but in the distance
an envious generality.

And not seeing what one doesn't want to see –
the contemporary leer and haircut
of the banjo player in the 'Shawm
and Sequence Band', a shadow
along the lip of someone axed
from memory by guilt,
a world which appears to have learned nothing.

Distortion by halation
as if we were afloat on music –
twenty/twenty vision is a myth
of the lost leader, the Emperor
asleep in the mountains.

And on to and up to death,
not believed in until
established by a unison of facts –
she has proved mortal
and you have observed an angel
in the cemetery rain.

Change your glasses
and you can't find the pavement
under your feet. Another change
and there are things in the prayerbook
will shock your pagan heart.

Making love and looking down
on a beautiful face,
the lips just pulling away

in absolute truth
from the apostolic teeth,
and know chillingly
there is no closer approach possible
than approximate ecstasy.

In seventy years one may learn to live
with defective vision. Take up art!
If your eyes meet Christ's eyes
they will never need to count
the limbs of the sleeping soldiers
in Piero's *Resurrection*.

Believing is not seeing
but a theology of doubt.
In dreams the exaggerations of language
become shapes and outlines
and at last the rower
thrusting the damned back into the sooty water
with his paddle is not just Charon
or the quacking guide a Virgil
nor yourself someone with a pen,
but all the creatures of a smalltime childhood
crowd to the boat and are classic
in the clearness, and of all greetings
none is so sharply sounded
from such known features
as the Mother's, out of sight of her son,
looking for the love which she was promised.

What I Have Written I Have Written

It is the little stone of unhappiness
which I keep with me. I had it as a child
and put it in a drawer. There came
a heap of paper to put beside it,
letters, poems, a brittle dust
of affection, sallowed by memory.

Aphorisms came. Not evil, but
the competition of two goods
brings you to the darkened room.
I gave the stone to a woman
and it glowed. I set my mind
to hydraulic work, lifting words
from their swamp. In the light from the stone
her face was bloated. When she died
the stone returned to me, a present
from reality. The two goods
were still contending. From wading pools
the children grew to darken
gardens with their shadows. Duty
is better than love, it suffers no betrayal.

Beginning again, I notice
I have less breath but the joining
is more golden. There is a long way to go,
among gardens and alarms,
after-dinner sleeps peopled by toads
and all the cries of childhood.
Someone comes to say my name
has been removed from the Honourable
Company of Scribes. Books in the room
turn their backs on me.

Old age will be the stone and me together.
I have become used to its weight

in my pocket and my brain.
To move it from lining to lining
like Beckett's tramp,
to modulate it to the major
or throw it at the public –
all is of no avail. But I'll add
to the songs of the stone. These words
I take from my religious instruction,
complete responsibility –
let them be entered in the record,
What I have written I have written.

The Killing Ground

Who comes in this disputed territory?
Just now the bored cat prowling
stops for a minute by the milk bowl
on the yellow, watching mat –
the kings of the earth rise up
and the rulers take breakfast together.
Our great men grant us equality,
they pardon us our lack of power
once they have considered
that we too die like them.

But this not yet. The air is lined
with demarcations of despair.
Each burly mote can tell
how she sat there at last breakfast,
night fumes in her dressing-gown,
and how she said the sun upon the square
was a massacre if only we would see
the bodies. Then the cats walked straight to her
with authórity from the tomb,
debating exits with their excellent tails.

Ever since that day there have been parties
of chomping sightseers to show round:
we want to know our nation's starting point.
Such pharaonic air, kippered happiness
which makes the postcards curl –
nobody believes the screaming guest
who says he sees a coopered figure
still at home. This cursed spot
brings thunder to the calm keyboard.

Be welcome then you cats who pass unscathed
through danger. You see me come to greet
friendship with apologies, knowing
I have never shifted from the killing ground.

The Imperfection of the World

It was after the idea of perfection
that the idea of reconciliation came –
nothing could mollify the one
and a sense of homecoming
justified the other.

Yet I foresee a new quadrangle,
ivy covering glazed brick,
and a man standing on a grille
while his colleagues decide
whether to stone him.

Or, to look less far, someone
in an apron discovering that love
is not what she wanted from life
but the name of the discontent
she feels in herself.

That the dreams which the gods bring
in their plural jackets are to show us
our ends cannot be disputed,
nor is the shape we make bearable
entering silence.

Yet we are haunted by our memory
of perfection, of setting out among ferns
for our father, the birds of the air
moderating painful noon
with their clamant cries.

Not even regret may stay in Eden.
Bellini's melodies are spoiled
by scratches on a record's surface
and a baby wakes to light fleeing
the face of nothing.

‘A bee is a device invented by a gene
to make more genes like itself.’
Therefore in the night when I cry
that I have been deserted, I am
once more made perfect.

The Future

It is always morning in the big room
but the inhabitants are very old.
Crooking her finger on a watering-can,
a precise figure of regret, no wisp
of her silver hair disturbed, drips succour
on a cat-predated plant. Words here
are shredded like its silver leaves,
they are epitomes of chanciness,
none will get you through the day.
When the sun fills the windows with its
misleading call to truth, the old woman
changes to a young girl, then to a man
from a novel looking up to ask
why things have gone so very wrong.
I am allowed, as if this were a dream,
to join them on their tableau.
We do not die, they say, but harden
into frescoes. This is what the future means,
her seeking me on her knees, poignant
as a phrase from a Victorian novel
or farewell spoken beyond a watercourse,
lyrical erotica I have no talent for –
Just the one room brightening, to which
hasten all the relatives of insecurity,
talking of my brief Bohemian days:
To be poised as the long-necked swan
or collared badger while the work
of worldliness is done, to stay the same
after the sun has gone, waiting merely
for light to show us up; the future
is to stand still with one gesture held,
a white glove entering a confluence.

The Story of Jason

As with all good stories, one cannot tell
If it is an allegory. If it is,
What should we do tomorrow,
In another week with a Thursday,
To redeem our ordinariness or find
Some virtue to take the place of courage?
But it is a quest with pictures
And we can set a ship like a barque
Out of literature or an Adriatic liner
Somewhere in the immanent foreground,
A view of the Symplegades behind;
Caparisoned horses, gods and kings
And a great deal of that burning saffron
Which came in with the fifteenth century –
This will suffice to identify
Jason and help him steal the fleece.

But then the sadness of pictures hits us:
We are far into the twentieth century
And must sit in innocence
Admitting that hell is possible.
We see the hero setting out
From an impoverished quay, his
Companions smiling conquistadores.
What we need are stories. Regard this
Poem up to now as preparation.
A certain loquacity is natural,
Long and high visions have prevailed.
Someone is alighting on the mole
And asking for the magic gardens.
He seeks old age and academies,
Wife and amanuensis ready,
Twilight around him and a bell
Ringing in the Palazzo Vendramin.

Garden of Earthly Delights

It has its corner shops, no dogs allowed,
Its tireless, unenfranchised natives of Mauritius;
Here too are city beehives, well endowed
By walks of public flowers; it knows ambitious
Men with calm rosettes, ignorant of history,
Their world a new place to be conquered
Because they have arrived in it; the sea
Far off is sobbing with its cargoes, word
Made flesh in geodesic domes, lithe coils
Of dog turd smoking in autumnal streets.
Its many mad show dream-syntax rules.
Not to live in the real world and other defeats
Are topics for the kebab clique. The city regards itself
In terror as in harmony. I invented Man,
It says, I named him Ghibelline and Guelf
And let his blood dry on my open plan.
There were far worse terrors to be fleeing from –
Moonscapes of imagining, plains where faces
Cut diamonds, a mirror is a bomb,
Pelting pastoral of nomadic races.
Our bourgeois duty is to make gods,
Expensive ghosts to act for sex; our cranes
Outsoar the trees for flowering rods,
We publish Nature with the trails of planes.
This turmoil is to fund a sense of loss:
Look up the legs of the garden-sunning nymph,
An angel sleeping with sword across –
It is the city-building goddess on her plinth.
Into this concrete garden come rumours of streams,
Gothic forests unexplored, the dales of death.
Better to be here than be forced to dream
Back the delta world which gives us breath,
Fonder for humans to tell the upper globe
They will not be intimidated, though

They know their end, wires to each lobe
Fixed by God's technicians. Now go
Forward to the formal park, fanfares of
Hedges and the sparrow-eaved rotundas filled
With music. Pavilions of self-love
Adorn the grass: and here you must build.

Alcestis and the Poet

As the little blue-tongued lizard runs across
The floor and clambers on the cushions, so I
Have spent my life in your service. I have
Risen from beds of my own melancholy to grant
Your distress an audience, heard the chorus of self
Desert its lord to swell your tragedy. It wasn't
Self-effacement but a bonding-up of time. We
Start with bodies from our wounded parents, not knowing
That the early flesh is useless, that its greyness
Towards death is what we love in it. So,
As young Shakespearean gestures, we glow among our feelings
And are pointless. Then, as the shades of madness
Intervene, we become important. Voices singing German ask
'Watchman, what of the night?'; geniuses ever upward tell
Of willing death, of lining tombs for study, Chattertons
Who persist in books. The soft arrival counts. Now,
When the rake of afternoon has laid the shadows,
We are ready to do each other service. Can
I march tongue-tied to the end; will you
Find the inexplicable, the out-of-reach-of-art
Intensity you mourn for outside Hades? I took
Your place and watched the stories grow. But it
Was no more than giving up a good position
In the queue – we are all for darkness. Death
Is in the small print, as Stevie Smith showed,
Ringing the word in galleys on her final bed –
Thus the loving woman does her duty and is
Woven into legend. The king sits in his kitchen,
Not certain if the world knows of her sacrifice,
Though don't his cats despise him? Are you there
Where each new disappointment makes you think that life
Is geared to reparation? And this time who is
Hercules? The joy of giving up, of saying sweetly,
'It could never have worked - real love must be
Thrown away or it will burn us.' The rest

Is timing. On the moon, they say, we find
The things we've sacrificed, pristine and waxing. Such dreams
Are cheats. Sited in great art, but tearful still,
The creatures that we are make little gestures, then
Go to nothing. The wind urges the trees to sigh
For us: it is not a small thing to die,
But looking back I see only the disappointed man
Casting words upon the page. Was it for this
I stepped out upon the stairs of death obediently?

Addio Senza Rancor

'Such past and reticence!' – George MacBeth

Two girls in their last year at school,
in the back row since they are taller,
stay young in the autograph album
which has slipped into sight from among
the fallen contents of the bookcase. Here are
the ingredients of sorrow, forever renewing
itself by generations – one was to die
at forty-one and the other at forty-four.

Not young by the standards of the world's unfairness,
only by those of our spoiled corner of it.
Why do we go on manufacturing misery,
waking when it cries, cleaning it for school,
clapping at the prize giving? The new girls
are in their mothers' clothes and the new fathers
stripping for the shining theatre instruments –
Unhappiness lives on, depression dies early.

Friends and lovers, kept apart by photographs,
we have made so much life to give away,
our generous faces must outlast us!
The shadows of that richness look over
my shoulder as I pick up a postcard
with a bent pin through it. Earliest yellow leaves
are appearing on the plane trees in the square –
the playground of maturity shall bury them.

Two friends high on death – what can I say to you,
not having experienced the mystery
which choked you? Nothing of the ordinariness
which lives in words and pictures trained you
for such priesthood. You are nowhere
in the evening light: what I see instead
are two white presences, playing with life,
smiling and letting it go without reproach.

Talking to You Afterwards

Does my voice sound strange? I am sitting
On a flat-roofed beach house watching lorikeets
Flip among the scribble-gums and banksias.

When I sat here last I was writing my *Exequy*,
Yet your death seems hardly further off. The wards
Of the world have none of the authority of an end.

If I wish to speak to you I shouldn't use verse:
Instead, our quarrel-words, those blisters between
Silences in the kitchen – your plainly brave

Assertion that life is improperly poisoned where
It should be hale: love, choice, the lasting
Of pleasure in days composed of chosen company,

Or, candidly, shitty luck in the people we cling to.
Bad luck lasts. I have it now as I suppose I had it
All along. I can make words baroque but not here.

Last evening I saw from the top of Mount Tinbeerwah
(How you would have hated that name if you'd heard it)
A plain of lakes and clearances and blue-green rinses,

Which spoke to me of Rubens in the National Gallery
Or even Patenir. The eyes that see into Australia
Are, after all, European eyes, even those Nationalist

Firing slits, or the big mooey pools of subsidised
Painters. It's odd that my desire to talk to you
Should be so heart-rending in this gratuitous exile.

You believed in my talent – at least, that I had as much
As anyone of a commodity you thought puerile
Beside the pain of prose. We exchanged so few letters

Being together so much. We both knew Chekov on marriage.
The unforgiveable words are somewhere in a frozen space
Of limbo. I will swallow all of them in penance.

That's a grovel. Better to entertain your lover with sketches
And gossip in a letter and be ever-ripe for death.
You loved Carrington as you could never love yourself.

I think I am coming within earshot. Each night
I dream comic improvements on death – 'Still alive
In Catatonia', but that's no laughing matter!

Perhaps I had Australia in me and you thought
Its dreadful health was your appointed accuser –
The Adversary assumes strange shapes and accents.

And I know, squinting at a meat-eating bird
Attempting an approach to a tit-bit close to me,
That our predatoriness is shut down only by death,

And that there are no second chances in a universe
Which must get on with the business of living,
With only children for friends and memories of love.

But you are luckier than me, not having to shine
When you are called to the party of the world. The betrayals
Are garrulous and here comes death as talkative as ever.

The Werther Level

Then he must wear his suffering like a sailor-suit,
something his mother found at a jumble sale
when she was being careful about money
and spending even more of it than usual,
what I mean is you can't do anything about clothes
unless you've got style – that is
flair which goes with feeling this is your world,
it fits you and there is certainly a lot
to be grateful for among the origins of love,
the trays of satisfactions.
I suggest that the mystery
is what is so daunting eventually. Over lunch,
an old friend tells you all the philosophies
or what they come down to, that another girl
will become available when this one's hysterics
have been outfaced, that you don't even look so bad,
in fact she'd go so far as to say that your body
at fifty is a better prospect than it was
in its tied-up, fitful twenties, and loneliness
is a sort of relief beside the obligatory togetherness
of parenthood – oh and so much more
and sincerely meant. But it won't work;
that rather well-dressed figure from the world's
most famous novel of self-pity hovers in sight,
so unlike today's casual narcissists, a pure
cavalier of auto-angst.
He has a message for you
which he puts with deliberate tactful insistence
as if through his speaking the pale certitude
of letter-writing in the great age could be reborn
and morning desks might once more
shine behind every window whose debates
consume the course of unkind love
as surely as the sun its own ascent.
'Do not think to escape responsibility

for living. Though the crime is not gazetted
punishment is palpable and that's the same
as law. Why have you never grown up,
grown out of that childish whimper brought back
from school, "it's so unfair." Try to think
of this flesh as clothing, not as envelope;
now we must do what is required of us,
write death into our Briefe, a discipline
of being garrulous for burghers. I light the lamp,
the room is dark in fits, as I sit to a piece
of paper white with wonderment. These
are the right clothes to welcome love in,
blood from a pistol wound or some trimming
from the newer Germany, a serious sight
among creatures light enough to miss the net of death.'

The Unlucky Christ

Wherever they put down roots
he will be there, the Master-Haunter
who is our sample and our
would-be deliverer. Argue this –
there were men before him,
as there were dreams before events,
as there is (or perhaps is not)
conservation of energy. So he
is out of time but once stopped here
in time. What I am thinking
may be blasphemy, that I
am like him, one who cannot
let go of unhappiness, who has
come closer to him through suffering
and loathes the idea. The ego now,
that must be like a ministry,
the sense of being chosen among men
to be acquainted with grief!
Why not celebrate instead
the wayside cactus which enriches
the air with a small pink flower,
a lovely gift to formalists?
Some people can take straight off
from everyday selfishness to
the mystical, but the vague shape
of the Professional Sorrower
seems to interpose when I try
such transport. The stone had to roll
and the cerements sit up
because he would have poisoned
the world. It has been almost possible
to get through this poem without writing
the word death. The smallest
of our horrors. When they saw him
again upon the road, at least they knew

that the task of misery would be
explained, the evangelical duty
properly underlined. Tell them
about bad luck, he said,
how people who get close to you
want to walk out on you,
tell them they may meet one person
even more shrouded than themselves.
Jesus's message at Pentecost
sounded as our news always does,
that there is eloquence and decency,
but as for happiness,
it is involuntary like hell.

Two for the Price of One

But the gift total is the gift in small
and before a structural horror of white paper
the maker finds a phrase which grows into
a power of everything. He cannot call back death,
he has licensed disappointment and regret,
his stay must be all imagination
and the working-out of time. Planes circling
the sky's rotunda may come down anywhere,
the Carmine frescoes crack before the eyes
of light machines. Magnificent and terrestrial things
are with the Spinners: some they leave in uncleaned
cupboards, some rise resistless like the sun.

★

That many of the Ark's denizens could not now report
to a new Deucalion is not the nastiest story.

Some living their lives in sulphuretted half-light
or slatted over shit might prefer extinction,

might watch enviously the passenger-pigeon
practise its lack of adaptation to the gun.

Survival can kill. Not just of cat and mouse and dog
but the million-faceted virus. Inventive Man

breeds a more economical red-meat quadruped
or white-fleshed avian source of protein

for his Nursing Homes, and hears above his hi-fi'd Bach
a cry from the drawing-board, the Unicorn's love laugh.

Lip Service

Game, set and match to the blubbering king.
What you do with lips is amazing
But you may prefer to eat, win points or sing.

They go over the side of the rubber raft,
All those losers, those words like *bereft*.
To come out of our warm books would be daft.

Heart Failure! The lower lip has whitened
With a blue pump-line like the back-end
Of our fattest silkworm. So rich, so frightened!

These mouth the rituals which we say the eyes
Are the home of, soul-cities. An old wheeze
Of the invisible, not to know truth from lies.

Two armies meet and recoil, they roll up their fronts
On a smile: gravestones wettened like fonts.
Fafner skulks here who was a giant once.

Thus the tongue which must tell the old old story
Of those who lie on each other long and warmly
Is betrayed and discovered. The chorus: 'Orrore! Orrore!'

Pope's Carnations Knew Him

But they knew they were on duty, replacing
the Rose of Sharon and the lilies of the field
for a gardener who never put a foot wrong.

It was their duty to rhyme in colour,
to repeat their reds and pinks and shield
the English rose with their Italianate chiming.

He had such a way with the symmetry
of petals, he could make a flower yield
an epic from its one-day siege. His rows

of blooms had their grotesques but they
took the place of music. They bowed, they kneeled,
they curtseyed, and so stood up for prosody.

No wonder Smart learned from their expansive
hearts that they loved the ordered, the well-heeled
and ornate, the little poet with the giant stride.

Each gossipy morning he sniffed their centres
and they saw him: the lines of paradise revealed.
God make gardeners better nomenclators.

Nights at the Opera

Sympathy for Scarpia

If she should have all the beauty, passion, honour,
Her lover the artistic temperament and courage,
Then why not Scarpia the will, the lust, the power –
All three are to get the crude justice of death.

Lucia in the Sky with Diamonds

Mad as she is, this girl
can't have the last word.
Her lover, equipped by the Scottish
Tourist Board, is going to bring
Italian opera into church. While she
encroaches on the constellations,
he extols a carnal love
or sepulchre of kisses wearing kilts.
The same contrivances evading death
honour a tone-deaf poet's end.

Waiting for Isolde

Love's wounds are transformation scenes.
The seagull over unarriving waters
swoops on the darkness of tumescence.
A relative named Amfortas elsewhere
moans in sympathy. If she never comes
he will grow old adoring Haydn.

How would you like to be called Wurm?

Villains are pillars of the ancien régime:
They have met innocence in a bad dream.

Papageno's Panti-hose

While Tamino learns philosophy
and how to play the trombone,
our good Viennese works to keep
his partner pregnant. Must it

come down to cafe society? Will
Papagena sympathise with all
Pamina's plaints? The opera
is forever fresh but the world
takes the waters out at Baden
and only half-reads letters.

If you can't join them, beat them

With that 'dumme knab' Siegfried around
Fafner had to be a sluggish dragon,
Wotan a quenchless ancient
and Brünnhilde in the Red Brigades.
Such things happen when fate comes between
two peerless lovers in their self-absorption –
the universe ends up a noisy joke.

O.K., Nerone

No opera ends less morally,
not even *Turandot*. The ruthlessness
of the young composer, which went with idealism,
now in old age becomes visionary:
only power lives on equal terms with beauty.

Audience at La Scala

Taught to suck from capitalist sores their venom,
Marxist First Nighters dress in well-cut denim.

Es sucht der Bruder seine Brüder

Faithful love has shown what it can do,
The syncopation gets more complex,
A cyclorama of celestial blue
Lifts our minds forever above sex.

About the New

Then this is the name
whispered in the street of horror,
(forgetting that we live just one time),
to try to bring to a point
the lines before your face at waking.

Yet it dirties the same paper
and cries out similarly
when oxygen is denied:
insipid over-achievers
worship it: it makes them profound.

There is Mr M
who has a spit problem,
grilling all writing but his own.
He has his generosity too,
he serves the cottage industry of words.

And think of the incomparable S,
putting away for Stratford,
for 'the dark house and the detested wife',
and afterwards through love
turning a sonnet into slippers.

A frivolous essay in aesthetics!
When they found themselves
in sordid cities where armless
pregnant girls begged for bread,
they cried 'Behold Neopolis!'

Do not confuse this with despair.
At any moment someone is born
wishing to forget everything
the world has ever known,
to found a university of starting.

I prefer the vision of Walter Pater
who didn't know much art
but looked at pictures to save them
from the scholars. God too
glances at our biros enthusiastically.

What an amazing thing, the confidence
of the world! It might be
'an accompaniment to a film scene',
'a systematically deployed semantic tic',
or just a dead sheep on the Downs.

Here at the end I must admit
that some things are definitely new
though undoubtedly made up
of bits of old things. To those in heaven
I shall say in greeting, 'Hello, you two.'

Which makes this a short essay
if rather a long poem by today's standards.
It took a few risks with syntax
but isn't innovatory. I expect to find
what's new beyond the encroaching fire.

About on the Serchio

For Ronald Ewart

Shelley's unfinished poem
must have been written near the mouth
on the flat dull stretch to Pisa.

Here, by the Devil's Bridge,
a glint on water is pollution,
though the heaped-up stones hide anglers
and hang-gliders float
impenitently down.

I am used to Tuscany
but not to the Garfagnana,
to the Beatles from a top piazza,
to SKYLAB and the high green figs.

There is no cure for the eye
and its pronouns
unless unhappiness be starved
like saints out of their country
of memory among the chestnut groves.

For Sophonisba Anguisciola

So much going for you,
a woman painter of the sixteenth century,
dying at ninety-seven
if the reference books are right,
and with such an elegantly
unpronounceable name.

It seems an attainment of grace
just to honour you and to forget
some undistinguished pictures,
almost as if the procession of geniuses
with their 'Triumphs of Time'
and their 'Feasts of the Gods'
passed right through the town
and out into the countryside
leaving you, me, my typewriter
and an honoured calling
to represent the human race.

Pienza Seen by Prudes

There is so much which poetry turns its back on,
The Rout of the Past, the you and you and you
For whom I don't exist, the crossing
Of these hills in our over-powerful car,
Up and down the fawn of Tuscany
To the Pope's town: clouds sail to worlds
Beyond us as we motor into visions
Harder than paint. Scattered by tyres,
Angels disperse to fresco-bearing trees.

The mind is made of Guide Books, factitious
Chapters of a biased history. Where local boy
Made good, things stay looking good, dust sheets
Over faction, and deracinated ankles
Swell on the way to Calvary. A little
Renaissance is put in the palm of hand
To keep the wonder venial. Today
Our poets are not fit to be provincial
Governors, nor will they fruit like olives.

The town has made a sculpture of the sky.
Pale prudes of their own blood approach
This vine-upholding vale looking for
Simplicities of everything too difficult.
Why, when the grandest of us little men
Is whisked away to Heaven, should survivors
Flounce to the parapet explaining things?
Sausages and wine are placed before us,
The wheel of work rolls past the perfect town.

At Lake Massaciuccoli

'Ecco il lago Massaciuccoli
tanto ricco di cacciagione
quanto misero d'ispirazione' – D'Annunzio

A huge bombardment on the lake's long plain
As green worlds collide and skim above
The oily surface – visible to us only
As a dust of spume and green confetti
Where small frogs jack-knife on to lily-pads –
Tall rushes begin beyond the rotting jetty
And over their grave heads an oriental bridge
Leads nowhere. Toffee-coloured heat
Holds the outdoor cafe and the pampered villas,
A stain of rice-fields in the middle distance –
Indiscrete lemons lean across the road
To naturalise the noonday tide of cars:
Italy still fights its history
With engines. Where, though, I ask myself
Are the descendants of those ducks Puccini shot
With all the skill of a Ferrari engineer,
Where the ghost of that armed man wading
'To terrorise the palmipeds of his adoration?'
Boom. Boom. Fall of the executioner's axe,
The cancer surgeon's scalpel, the gong
Which announces that death's challenge
Has been taken up. Eighty cigarettes a day –
Pilgrims waiting at the gates observe
The lung-coloured lake. *L'homme armé*
Goes too far back and yet walled Lucca
Has a league of high composers no less
Pungent than Castruccio. Putting on his waders,
He might think of art, of facing the public
Armed with the visible part of dreams. Disappointment,
For all his calculation to a quaver's whisper,
Leaves him no resort but slaughtering ducks.

No one produces the art he wants to,
Everything that he makes is code,
To be read for its immaculate intention.
Then in death he finds the final disappointment,
That no clarity comes anywhere, the perfect
Vision has gone into the mist, as when dawn
Wakens the wet-winged skimmers on the lake
And every hazy lineament lures the hunter
Into a picture-postcard world. *O mors inevitabilis,*
Not to be held back by more than function,
A pot of Stephens' Blue Black Ink, a gale
All night among the pines and yet no air
Upon our planet – nothing so well observed
As pain, apotheosis of things out of place.
To return then after some small adultery
To the mystery of fiction; to write letters
To the world's four corners while mosquitoes
Shake like stage scrim across the door:
There must be a vision, perhaps of cruelty
In Venetian Peking (better at least
Than the sort of thing D'Annunzio would offer) –
It hardly matters, since the big tunes
Wait in the desk for him to pick them up
And a wife can keep one's view of sorrow fresh.
He should think himself most fortunate
Never having needed to be autobiographical.
At the lake's side, I too, maker of this
Near dramatic monologue, honour him truly
Yet could not bear an enlargement of
The world these frogs and teeming spawn inhabit.
He was not so soft: what he saw
Was this lake made into the world – not to be
Changed or pitied but crying through the night
Abandoning life for love. The dark will come
To every average denizen the same,
Sounds upon the shore staying for waftage.

The Unfortunate Isles

From the gunwale the dazzle is like spray,
yet past the low reefs and their morning surge
gloriettes of cacti wait, summer houses
flensed from trees, a palimpsest
of childhood books and hardihood
of castaways. An hour ashore
and you forget your face, a week later
your colonists are raising ghosts
of thc great world. Can all these
have swum in from the wreck – dreamers
stung by lexicons, lackland dowsers
imposing on the dark, sorters of bird droppings
granting interviews by gravesides?
Open the dictionary of discontinuity
and read about these islands. ‘Over the swell
of everyday begins that archipelago
where the irreligious may taste ecstasy.
This is at balanced armslength from
the Heroes’ Home, and at any time a battlefield
is fulcrum. Follow your face here
among the scone-and-strawberry pantiles,
see a dream identity turn a corner,
aunts and uncles of the might-have-been
distributing their malaise. All currents swing
us to these shoals and as we pass
we note imagination is bureaucratised,
each island so characteristic of itself
we are no more than sorted as we float
like seeds to a soft imprisoning.
Here collects the melancholy
of deep coincidence, an airborne spored
unhappiness which centuries have used
to establish hallmarks. Europe, Asia,
lost Atlantis are hardly continents
seen beside the wreck of self – there stands

the Principality of Childhood reduced
to a crumpled letter, there a rain tank
rusting into canna flowers which marks
the courtliness of love. Nobody weeps here
for what he's lost, since everything is home.
Each is a creature calming himself
with more anxiety. The prevailing wind
blows memory in your face, and up the beach
the harmonies of death return to breed.'

Landscape with Orpheus

'Man lebt nur einmal, dies sei dir genug.'

It was as if the film had stuck, he was always
Back at the point where he moved up the latch
And stood facing down the street, aware of
The cicadas turning themselves on in both tall
And dumpy trees: what he saw was limited
But included lakes of dirt-in-asphalt
Before his feet, the unfortunate slug about
To cross the pavement with no more instinctive
Knowledge of its danger than he had of the sun
Perhaps on his neck, and of course always
The Dutchman's Pipe flowers he never failed
To notice, their purple mouthpieces like Disney
Saxophones, edible, sexual and howling for the dead.

It would take a lifetime to make it to the ferry,
A sunstroke's distance amid pavilioned leaves
And so desirable an ending. Well, there was a life
To spend and this was time, the softest element,
Like sap from poinsettia leaves, the milky pus
Of dreams – Eyes stood on tiptoes in those hedges,
So perhaps he should begin. It was late and it was early
In his sorrow and he had the world's tunes to play
And a landscape of peace and obsession there –
To see it all stretched out and hardly a step taken,
Such was the gift of time, walking down to the ferry
With love to come and snake-bite and the bitches flying
As calm as tapestry, in light-soaked Poussin shades.

Praise of his bloodstream flowed on then in sounds.
That this untrained imagination out of mercantile
Forbears should be Emperor of Cadences didn't surprise,
Don't we all know we are immaculate in our dress
Of self, and the twenty billion succinct souls

Hanging on God are just light in the distance
By which pilgrim feet find tracks to follow,
And that this fold of fact would undo and show
A hidden nothing if we blinked? The place of the ordinary
Is on the throne: save afternoons for judgment
And every morning for our table music. Could he have
Passed the big house with the haunted windows
And nipple-pointed fence, he had hardly moved?

But the stickiness underfoot was disquieting,
Perhaps the land was Avernus, with those
Bamboo raggednesses above the fence and the pale smell
Of warm tar on the air. Through fur of sugar-grass
He saw the river and all remote existence
Sculling across the darkened tide. What blew in his face
Were words, those he would speak in love and those
Which fattened on betrayal. The words for death
Were still unknown and yet he knew they sought him
On the street. A wind of big mothers mixing drinks
Caught him suddenly with laughter. What if he got
To the wharf, what if the ferry with a Lady's Name
Were there? He sang, in case, 'Goodnight, deceiving world!'

The cicadas stopped. Silence grew into a theatre
With everybody watching. According to the small print,
When love has failed to come you choose your end
By divination. Child or old man, now is the hour
And memory's prevarication cannot last.
With final breath whisper us your Eins, Zwei, Drei.
The sun in intervention breaks the sky,
The camera is rewound and there is the old latch,
The gate, the pepperina tree, the ferry rounding
Onions Point. The future must be crowded into now,
Paradise and hell on deck. Viewed through the telescope,
The Town Hall clock shows Orpheus looking back.